save the . . .
TIGERS

by **Christine Taylor-Butler**
with an introduction
by **Chelsea Clinton**

PHILOMEL

PHILOMEL BOOKS
An imprint of Penguin Random House LLC, New York

First published in the United States of America by Philomel Books,
an imprint of Penguin Random House LLC, 2022

Text copyright © 2022 by Chelsea Clinton

Philomel Books is a registered trademark of Penguin Random House LLC.

Visit us online at penguinrandomhouse.com.

Library of Congress Cataloging-in-Publication Data is available.

Printed in the United States of America

ISBN 9780593404201 (hardcover)
ISBN 9780593404218 (paperback)

10 9 8 7 6 5 4 3 2 1

WOR

Edited by Jill Santopolo and Talia Benamy
Design by Lily Qian
Text set in Calisto MT Pro

*Dedicated to all the kids
who love this planet as much as I do.
Together we can make a difference.*

save the . . .

save the . . .
BLUE WHALES

save the . . .
ELEPHANTS

save the . . .
FROGS

save the . . .
GIRAFFES

save the . . .
GORILLAS

save the . . .
LIONS

save the . . .
POLAR BEARS

save the . . .
TIGERS

save the . . .
WHALE SHARKS

Dear Reader,

When I was around your age, my favorite animals were dinosaurs and elephants. I wanted to know everything I could about triceratopses, stegosauruses and other dinosaurs that had roamed our earth millions of years ago. Elephants, though, captured my curiosity and my heart. The more I learned about the largest animals on land today, the more I wanted to do to help keep them and other endangered species safe forever.

So I joined organizations working around the world to support endangered species and went to our local zoo to learn more about conservation efforts close to home (thanks to my parents and grandparents). I tried to learn as much as I could about how we can ensure animals and plants don't go extinct like the dinosaurs, especially since it's the choices that we're making that pose the greatest threat to their lives today.

The choices we make don't have to be huge to make

a real difference. When I was in elementary school, I used to cut up the plastic rings around six-packs of soda, glue them to brightly colored construction paper (purple was my favorite) and hand them out to whomever would take one in a one-girl campaign to raise awareness about the dangers that plastic six-pack rings posed to marine wildlife around the world. I learned about that from a book—*50 Simple Things Kids Can Do to Save the Earth*—which helped me understand that you're never too young to make a difference and that we all can change the world. I hope that this book will inform and inspire you to help save this and other endangered species. There are tens of thousands of species that are currently under threat, with more added every year. We have the power to save those species, and with your help, we can.

Sincerely,

Chelsea Clinton

save the . . .
TIGERS

CONTENTS

- - - - - - - - - - - - - - - - - - -

1

HOME ON THE RANGE

If you wanted to find a tiger, you might try looking in a rainforest. Or you might try a swamp. Of course you might also try looking in places with plenty of tall grass. A savanna would do quite nicely. Even a cold, frozen tundra would work. But even though tigers live in many climates, there is only one continent where you will find wild tigers. That continent is Asia.

Tigers in the Wild

How many tigers are still living in the wild? That's not an easy question to answer. Scientists think there were once one hundred thousand tigers living in Asia. That's enough tigers to fill every seat in the Roman Colosseum twice. Today there are probably fewer than four thousand tigers still living in the wild, so the seats would be almost empty. To find the tigers that remain, be prepared to travel. There are only six subspecies (or types) of tigers left on Earth. But they don't live in the same places.

Sumatran tigers live in tropical rainforests and mangrove swamps. Because Sumatra is an Indonesian island, these tigers don't ever come in contact with other types of tigers.

Indochinese tigers are found in the evergreen

A Sumatran tiger.

forests and mountains of Vietnam, Myanmar, Thailand, and southern China. There might be tigers in Cambodia and Laos, but no one has seen any recently, so we don't know for sure.

Malayan tigers live in dense, moist tropical forests and brush on a strip of land called the Malay Peninsula. It is surrounded by water on three sides and contains parts of Malaysia, Thailand, and Myanmar.

Siberian tigers (also known as Amur tigers) live in the north where it's often cold and snowy. They make their homes in Chinese and Russian pine forests.

Sadly there are only a few South China tigers left on Earth. To keep them from becoming extinct, the few that remained were first moved to Chinese zoos for breeding. But the tigers did not thrive at the zoos. No cubs were born. Now they are being relocated to captive reserves, where they'll live until they can be released into the wild. Combined, these five types of tigers are only a small fraction of the tigers still living in the wild. The rest live in India. In 2018, scientists counted almost three thousand Bengal tigers living in that country's forests and grasslands. That's three-quarters of all the wild tigers on the planet!

These tiger counts are estimates because scientists have found it difficult to count them. You may be wondering why. Well, Asia is gigantic! It contains 30 percent of all the land on our planet. At more than 17 million square miles, it's so big it could hold two North Americas— or five Europes. If you spread the tigers out across the continent, there would only be one every 4,250 square miles. So finding those tigers is not easy to do.

A Very Big Territory

But all that space isn't so bad for the tigers! They are solitary creatures that live and hunt alone on land called a territory or home range. A male tiger's home range overlaps with the home ranges of two or more female tigers. The male visits long enough to start a family then leaves to patrol his territory and hunt.

The size of each home range depends on how much food is available. If there are plenty of animals to hunt, the range might be as small as nineteen square miles, making those tigers easier to track.

If food is scarce, the range will be much larger. Siberian tigers have the largest home range, almost four thousand square miles. That's partly because there aren't a lot of people or cities to get in their way. Because their home range is so large, they might overlap with another home range. But you would rarely see two male tigers in the same place at the same time.

The right habitat provides shelter for raising a family. A female tiger, or tigress, raises her cubs by herself. To do this she creates a home called a den. The den might be in a cave or a hollow tree. Other times it will be in thick grass or a clump of small trees and shrubs. The den hides her cubs from predators (other animals that might harm them) while she is out hunting for food. If the den stops being safe, the tigress will use her mouth to grab each cub by the neck and carry them, one by one, to a new den.

A Siberian tigress carrying her cub by the neck.

So as you can see, finding a wild tiger is hard. They are constantly moving around to find shelter, to find safety and to find food!

Meat for Dinner

In a typical week, you probably eat a variety of food groups. That makes you an omnivore. Tigers are carnivores. Their bodies do not need the fruits or vegetables that people eat to stay healthy. Instead, they eat as much as eighty pounds of raw meat in a single night. Meat protein helps tigers build and maintain strong muscles.

How does a tiger get that much meat? They're apex predators. Apex predators are animals at the top of the food chain. They have few natural enemies.

Right: A tiger jumps high to catch prey.

When most tigers chase their prey (the animals they're going to eat), they can run as fast as forty miles per hour. That's faster than the speed of many cars on a city street. Siberian tigers are the fastest. They can run sixty miles per hour, which is about as fast as cars on some highways.

Even so, tigers weigh hundreds of pounds. All that weight makes it hard for them to run long distances. Instead, tigers are ambush hunters. They may hide behind a tree, a clump of tall grass or a group of large rocks until an animal passes by. They might follow their prey for up to a half hour. When the tiger gets close enough, it will sprint forward and pounce before the prey has a chance to escape. The tiger holds on to its prey's neck until it stops breathing. If other predators are nearby, the tiger will drag its prey to a safe place to keep from having to share the meal.

Not all tigers eat the same animals. Remember, Asia contains many different habitats. A tiger in Sumatra might eat fish, monkeys, wild boar or deer. A tiger in Thailand might eat antelope, turtles and pigs. A Malayan tiger might eat all of those things plus elephant and rhinoceros calves.

Unfortunately for the tigers, not all hunts are successful. Scientists estimate that a tiger catches prey once out of every twenty attempts. Those are not great odds for a hungry tiger. When prey is scarce, tigers may sometimes eat insects like termites. A pound of termites contains more fat and protein than a pound of steak. But termites are tiny. It would be impossible to eat enough of them to satisfy a tiger's appetite.

Have you ever heard someone say "man-eating tiger"? That's an unfair label. Tigers rarely attack humans unless they are sick and too weak to hunt for their normal prey. Tigers mostly avoid us. To be on the safe side, some people wear masks on the back of their heads or hats with eyes when walking through an area where a tiger has been seen. This tricks the tiger into thinking the humans are always facing it, so there's no opportunity to ambush them from behind.

Champion Swimmers

In addition to meat, tigers also need water, and the best home ranges have large sources of fresh water for them to drink from. Tigers are also very good swimmers. Their powerful muscles and big, strong paws allow them to

A Malayan tiger goes for a swim.

paddle long distances and catch fish without sinking. The webbing between their toes helps them to push the water away as they swim. If it is hot outside, water also helps them stay cool.

So be careful if you're in a river or pond in a tiger's home range—you may end up swimming with one of the most powerful cats in the world!

2

THE LARGEST CATS ON EARTH

Born Small but Grow Fast!

The scientific name for a tiger is *Panthera tigris*. They are the largest cats on planet Earth. Tigers don't start out big, but they grow fast. A tigress is pregnant for one hundred days before she gives birth to cubs. Newborn tiger cubs weigh two to four pounds when they enter the world. That's smaller than a newborn human baby, which weighs an average of seven pounds. But tiger cubs grow much more quickly. By the time the cubs are six months old, the males weigh about one hundred pounds. Females weigh about seventy pounds. Compare that to a human baby, which weighs an average of sixteen pounds at the same age.

For two months, the new cubs drink their mother's milk to survive. After that they can

Tiger cubs hiding in their den.

start eating meat. They're still too young to hunt on their own, but they can follow their mother to the hunting grounds and eat the prey she kills.

The cubs stay close to their mother until they are eighteen to twenty-four months old. Until then, they play with their brothers and sisters. They chase and pounce on each other. This helps them practice hunting and defending themselves. They'll need those skills to live on their own.

Female tigers create territories close to their mothers. They can start having their own cubs when they are three or four years old. Sadly, many young tigers live only a few years.

The males that survive leave to create their own territories too, and go even farther away. That can be dangerous if they go to an area a stronger tiger has already claimed.

As Big as a Car!

Most tigers grow between eight and ten feet long. That's about half the size of a large SUV. Males are usually bigger than females. Their weight varies by subspecies.

The smallest is the Sumatran tiger. An adult male weighs about 265 pounds, which is only twenty pounds more than the average football player. Malayan tigers are the second smallest, with males weighing about 280 pounds. South China tigers weigh as much as 330 pounds. Indochinese tigers weigh up to 400 pounds. That's the average weight of a Japanese sumo wrestler.

Bengal tigers are much bigger. They weigh as much as 480 pounds and grow almost ten feet long. That is almost three times heavier than an average human and almost twice the length.

But the biggest and longest of them all? Siberian tigers! They weigh as much as 660 pounds and can measure up to eleven feet.

Mammals Just like You!

Tigers are warm-blooded, just like you are. Your body can make its own heat, which lets you live in most climates (with the appropriate clothing, of course!). In cold weather you might put on a sweater or a coat. Tigers, on the other hand, have fur to keep them warm. The fur traps a layer of air, which keeps heat from escaping a tiger's body too quickly.

Siberian tigers are larger than other tigers, and their fur is longer and thicker. That helps protect them in harsh Russian winters.

Tigers that live in rainforests have fur that is shorter and not as thick as a Siberian tiger's fur. Even so, the fur still traps some of the heat they make. Those tigers cool off by panting the same way dogs do. This releases heat through their mouths. Between swimming and panting, tigers really do know how to stay cool!

No Two Tigers Look Alike

One of the first things people notice about a tiger is its beautiful stripes. Tigers have an average of 120 to 150 stripes on their bodies—mostly on their backs, sides, legs, tails and faces. The stripes help tigers hide in tall grasses and shrubs—even if the grass is green! That may not seem logical, but it's true. Many prey are color-blind. They can't tell the difference between colors like green and orange. The stripes on the tiger's fur blend in with the leaves and grass in its hunting grounds.

Another cool thing about tigers' stripes? No two tigers have exactly the same pattern. Their stripes are as unique as our fingerprints. And here's a fun fact: if you shaved a tiger's fur you would still see the stripes on its skin.

Orange, Yellow and More

If you wanted to draw a tiger, you might draw it with orange fur and black stripes. The orange color is created by a pigment called pheomelanin. The orange may be deep and dark, or it may be light and yellowish. But some tigers aren't orange at all!

A few Bengal tigers are born with white fur and brown stripes. Their bodies don't make the pigment needed to have orange or black fur. This is very rare. Only one out of every ten thousand Bengal tigers is born white. In 2012, only twelve white tigers were seen in the wild. None have been seen in recent years. White tigers are now only found in zoos and breeding programs.

Pigment color isn't the only difference in tigers. Orange tigers have amber eyes. White tigers have blue eyes.

Have you ever noticed the white spots on the backs of a tiger's ears? All tigers have them.

White spots are found on the back of every tiger's ears.

Scientists think the spots trick other animals into believing the tiger is looking at them when the tiger is facing the opposite direction. This might discourage a predator from trying to sneak up on the tiger. It might also help cubs see their mothers at night.

Bodies Built for Strength and Speed

You are a vertebrate. Tigers are vertebrates, too. That means you both have a strong backbone that supports your body. A tiger's collarbone is small, which makes it easier for them to take long steps. Their tails can be as long as forty-three inches. They're helpful for balancing when tigers are standing, turning or running.

Now take a look at your body. Do you notice that your legs are longer than your arms? The same is true of tigers. Their back legs are longer than their front legs. This helps them jump long distances. Tigers are so strong that they can leap forward as far as thirty-two feet. That's farther than any human has ever jumped (twenty-nine feet). So if you are being chased by a tiger, it is not likely you can outrun them—or out-jump them—even with a head start!

Try jumping high in the air. Did you make a sound when you landed? Even though tigers are heavy, they aren't noisy jumpers. The padding on their feet helps them land softly behind their prey. That's one reason why they are dangerous. If a tiger jumps behind you, chances are you wouldn't hear it.

Tiger paws have thick padding to protect their feet.

You wouldn't want a tiger to scratch you. Their claws are curved and can grow four inches long. Their front paws have five claws, with one farther back than the others. That helps a tiger climb something or hold on to prey. Their back paws have only four claws each. When tigers don't need to use their claws, they can pull them back into their paws to protect them.

Now try this. Move your jaw from side to side. A tiger can't do that. Their jaws are designed for power and only move up and down. If a tiger bites down on something, they are strong enough to hold on no matter what.

If you were brave enough to look inside a tiger's mouth you would see thirty teeth; most human adults have thirty-two. Like you, tiger cubs are born with temporary teeth. Over time,

A tiger's tongue is rough and sharp.

they are replaced with permanent teeth. Tigers have sixteen teeth on the top and fourteen teeth on the bottom. The two long front teeth are canines, which are good for biting and holding on to things. They can grow up to three inches long. Humans have canines too, but they are not long like a tiger's. Smaller teeth between the canines are called incisors. Tigers use their incisors to tear meat from a bone.

By the way, while you're looking into the tiger's mouth, you'll notice that its tongue is rough. It's covered with sharp ridges that help it shred meat. So you might want to avoid having a tiger lick you, even if the tiger is friendly.

Unusual Communicators

Tigers don't purr like house cats. Instead, they can roar so loudly it can be heard two miles away. Their roars have been measured at 110 decibels. That's louder than a jackhammer or a motorcycle. It's as loud as a rock concert and would hurt your ears if you stood too close.

The roar contains an unusual sound called infrasound. This sound is so low, humans can't hear it. Infrasound travels long distances and can confuse or paralyze other animals for a short time. Tigers also snarl, growl and moan. They might use these sounds to call their cubs. Or they might use them to warn predators to stay away.

Do you cross your arms when you're angry? Tigers swish their tails to show it. Beware! If a tiger arches its back and lowers to the ground, it may be planning to attack.

You might see a tiger scratching the ground or clawing a tree. That's a way to communicate the territory is already claimed. Special glands in a tiger's paws leave a scent on the ground or in the deep grooves it scratches into a tree.

Finally, tigers mark the ground with urine

and feces. The smell helps keep other tigers off their home range. It also helps cubs find their way back to their mothers if they wander off.

Active While You Are Sleeping

You are less likely to see a tiger hunting during the day. That's because tigers are nocturnal creatures. That means they sleep during the day and hunt at night. It takes very good eyesight to see in the dark. Tigers can see even better than humans after sundown.

Take a look at your eye. Do you see that black dot in the middle? That's your pupil. Pupils grow larger or smaller to change how much light enters your eyes. Tigers have pupils too, but they open much wider.

Inside your eyes are special cells called rods and cones. Cone cells help us see details when

Tigers sleep a lot during the day.

there is plenty of light. They help us tell colors apart. Rods can't see colors but are very good at seeing shapes and movement when it's dark. Tigers have more rod cells than cone cells in their eyes, which helps them hunt better at night.

Here's a cool fact: if you shine a flashlight at a tiger's eyes, they seem to glow in the dark.

Inside the tiger's eye is a special layer of tissue called the tapetum lucidum. That's Latin for "bright tapestry." Some scientists compare this layer to a mirror. It bounces more light in the tiger's eyes, helping it see better. Not every animal has a tapetum lucidum. For those that do, they're not all the same color. A tapetum lucidum might glow yellow in a cat, but green in a tiger. (You don't have that layer in your eyes, which is why your eyes don't glow in the dark.)

With such great eyesight, do tigers chase animals all night? Actually, no. Have you ever played a game of tag? It takes a lot of energy to chase your friends and catch them. Hunting is the same way for tigers. To rebuild their strength, tigers sleep as much as twenty hours each day. You might see them sleeping on rocks or in the grass. On a very hot day they will look for shady spots to rest. Zoos sometimes open after dark so you can see nocturnal animals when they are most active.

But here's a warning: don't sneak up on a tiger while it's sleeping. The whiskers on a tiger's face are so sensitive they can detect changes in the air around them. A tiger would know you were approaching even before it opened its eyes.

Now you know what makes tigers so powerful. You would think that would be enough to keep tigers safe. Unfortunately there is one predator that is even more dangerous than they are. It is the only predator responsible for decreasing the number of tigers on the planet.

That predator is us.

3

QUICKLY DISAPPEARING

For thousands of years, tigers have been a symbol of power, courage and strength. They're an important part of some Asian myths and legends. Tigers are also the national animal in some countries. Even so, the number of tigers left in the wild is low. If this is not changed, tigers will become extinct in some places. Scientists don't want that to happen—and we shouldn't either.

There are many organizations in the world working to save tigers. One of the largest is the International Union for Conservation of Nature (IUCN). They monitor the health of our planet and keep a list of living things that are in danger and need our attention. They call it the IUCN Red List of Threatened Species™. The IUCN has studied more than 138,000 species so far. Those that are threatened are grouped in seven levels of increasing danger:

Least Concern: These animals are doing well in their habitats, and their numbers are steady or increasing.

Near Threatened: These animals are safe for now, but there are signs they may be in trouble in the future.

Vulnerable: These animals are at risk of extinction, but the risk is still low.

Endangered: These animals are at high risk of extinction. Their numbers are falling, and they are losing large amounts of their habitats.

Critically Endangered: These animals have the highest risk of extinction in the wild. If we don't take action, these animals may disappear in the future.

Extinct in the Wild: The only places to find these animals are in captivity, such as in sanctuaries and zoos. They are no longer found in their natural habitats.

Extinct: There are no more animals of this type living on the planet.

Tigers are listed as Endangered on the Red List. That means that there are fewer of them than there used to be. The tiger population is slowly increasing, but not fast enough for tigers to be considered safe again. There are now about four wild tigers for every hundred that lived on the planet a century ago.

Those numbers don't tell the whole story. There were once nine tiger subspecies on our planet. Now there are only six. The Caspian, Javan and Bali tigers are gone forever.

You might ask yourself, "If tigers are so powerful, then why are they disappearing?" There are many reasons.

Poaching

Poaching is what happens when hunters kill or capture wildlife that is under legal protection. Some people hunt and poach tigers for their body parts and beautiful skins. Those are sold in stores and markets around the world. In some cultures, the bones are especially popular for soups and traditional medicines. Some people believe tiger bones will give them a long life.

This tiger skin was turned into a rug.

People working to save tigers are trying to stop the sale of furs or medicines and foods made from tigers. But that doesn't always work. In some countries, the money a tiger's body can bring might be more than many years of salary earned in another job. Tiger pelts have been sold for up to $20,000. A rug made from a tiger pelt was once advertised for $124,000.

China buys more tiger parts than any place in the world. In 2013, many organizations convinced the Chinese government to stop this practice. Five years later, China changed its mind and allowed citizens to use tiger bones in medicines again. Tiger bodies could be used if the tiger was born in captivity. This made things worse for tigers everywhere. Poachers began hunting more tigers in other countries. Tiger farms bred more tigers in order to sell the bodies. It is one of the reasons tigers are still on the Endangered list.

Killing for Sport

Not everyone hunts tigers for money. Some people do it for fun. Those hunters like the idea of killing an apex predator. Having a tiger as a trophy gives them something to show off to other people. Because weapons now exist that make it possible to hunt tigers from far away, more hunters feel safe doing this. But many people don't like the idea of killing a tiger for fun or pride, especially with so few left on the planet.

It's especially troubling when hunters kill a tigress. That makes it harder for the tiger population to recover. If a tigress is killed, her cubs may not survive. Other tigers will not adopt or feed them. Killing a tigress also means fewer cubs will be born in the future.

Disappearing Habitat

In addition to hunting, there are other reasons tigers are endangered. Imagine if someone took away your house and the places you shop for food. It would be hard to survive.

That's happening to tigers and their habitats. Over time they have lost 95 percent of their territories for many reasons. Two of the biggest ones are:

Deforestation

Deforestation happens when humans cut down trees for fuel, homes and other uses. Southeast Asia, for example, lost more than 235,000 square miles of forest between 2001 and 2019. That's more land than the entire state of California. Remember, each tiger needs hundreds of square miles for hunting. Deforestation not only takes away a tiger's habitat, it also destroys the habitat that a tiger's prey lives in as well. If tigers can't find enough animals to hunt, their numbers will continue to drop until there are no more tigers left in the wild.

Farming

Claiming land for farms is another way tigers lose their habitats. As more people are born, more land is needed to support them. It is not only land to build houses or cities, but to grow crops. Yet more land is taken so farm animals can graze. When tigers have nowhere to go, they will go where the humans are. That's a recipe for disaster. For tigers, a farm animal looks like prey on its hunting grounds. But the farmer sees tigers as a dangerous predator. Farmers may hunt and kill tigers to protect their farms and livestock.

Climate Change

Our planet is changing. More specifically, our climate is changing. Climate is the type of weather you find in a specific place on Earth over time. For instance, we know that the North and South Poles of the earth are cold. The equator, the invisible line around the center of the earth, is hot and humid. You've already learned that tigers have adapted to the climate where they are born. But now the climate on Earth is changing faster than they can adapt. And it's shrinking the tiger's habitat, too. Here's how:

Flooding

Because the earth is getting warmer, glaciers and icebergs in polar regions are melting. Snow and ice found on the tops of mountains are also melting. The water flows into waterways and oceans. This makes them rise and flood habitats where tigers live. For example, rising water is carving away the mangrove forests used by Bengal tigers.

Even worse, tigers, like humans, need fresh water to drink. They cannot drink saltwater because too much salt in their bodies makes them very sick. When ocean water floods the land and mixes with the fresh water sources tigers use, the tigers have to find new places to live.

Forest Fires

Climate change also increases the chances of natural disasters like forest fires. Many forests on Earth have seasons in which fires are a normal part of their life cycle. But fires there are getting worse than they used to be, and there are places on Earth where forest fires are not normal. Arctic climates in Russia and China are cold and have plenty of moisture to help trees to grow. Hotter, drier weather dries out the pine trees that grow there and makes it easier for them to burn. Fir and spruce trees grow in their place. Those trees do not create the best habitat for the animals that Siberian tigers need to hunt. In 2021, Siberian wildfires burned more land than the total of all the other fires around the world that year.

What is happening on Sumatra is another example of this problem. You don't expect fires in a humid rainforest. But in August and September of 2019, that's exactly what happened. Fires destroyed almost a third of the tiger habitat in Indonesia's Sembilang National Park. Another part of the park was destroyed two months later. Because the island is surrounded by water, Sumatran tigers can't leave or make new territories on the mainland. Those tigers are forced to move to another part of the island. They may have to fight or kill other tigers to claim territory on the land that remains. This is one of the reasons there are less than four hundred Sumatran tigers alive today.

The Palm Oil Problem

You may not have heard of palm oil, but you have probably used it—or even eaten it! Palm oil can be found in everything from soaps and lotions to the cookies and cakes you buy in grocery stores. Half of the palm oil created is used to replace air-polluting fossil fuels like the gas that runs a car. You might ask, "What does this have to do with climate change?" The answer is quite a lot.

The main problem is that palm oil comes from oil palm trees. In order to grow oil palms, traditional rainforests are destroyed to make

A forest in South Asia was burned to make room for a palm oil plantation.

room for them. For example, the island of Sumatra has lost 85 percent of its forest in the last fifty years just to grow oil palms.

Dr. Michael Coe, a scientist at Woodwell Climate Research Center in Massachusetts, called rainforests "a giant air conditioner" for the earth. So while many people think they are saving the earth by using palm oil, they are actually hurting the rainforests that help control climate change and destroying habitats used by tigers and their prey.

Living in Captivity

Sometimes people trying to help tigers wind up being part of the problem. They capture them or buy them from other people and put them in petting zoos and animal attractions. Some keep tigers as pets. Scientists estimate there are as many as thirteen to fourteen thousand tigers in captivity. That's almost five times more tigers than are found in the wild.

That might sound good—lots of tigers is the goal, right? Not exactly! Many attractions are illegal. Imagine if you were not allowed to walk to school or go to a playground. Instead, you were locked in a tiny bedroom for weeks and months and years while people paid money to stare at you and feed you. Pretty awful, right? It's similar to how many of these tigers are treated.

Here's a good example. Until 2016, 147 tigers were being kept at the famous Tiger Temple in Thailand. Tourists paid as much as $200 or more to hold tiger cubs and feed them with baby bottles. They took photos with larger tigers and walked them on leashes. The owners made a lot of money, so they didn't want to stop. But the tigers were not fed properly and were often beaten. Instead of having large territories to hunt, tigers were kept in small concrete cages. Many had serious medical problems. In 2016, the Thai government rescued them and brought them to wildlife sanctuaries. Even

*A tourist poses with a tiger
at the Tiger Temple in Thailand.*

though they were safe, more than half of the tigers soon died from illnesses they got while living at the temple.

Asia is not the only place where illegal tiger trading is a problem. It happens in the United States, too. In 2007, the Captive Wildlife Safety Act tried to fix that. The new law made it illegal to sell or move big cats across state lines. Even so, people kept doing it in secret. Why? Because the federal law doesn't ban what happens inside a state's boundaries. Some states created laws making it illegal to own tigers as pets. But people could still sell or trade them as long as the tigers weren't moved to another state. Alabama, Nevada, North Carolina and Wisconsin have no laws to protect tigers at all. People in those states can keep one in their backyard!

Here's another example of how keeping tigers in captivity is more harmful than helpful. Have you ever visited a roadside attraction with animals? They are not licensed zoos and are often unsafe. Just like the Tiger Temple, these attractions are successful because tiger cubs are cute and attract a lot of tourists. But cubs grow quickly, and in a few months they become too dangerous for tourists to handle. The solution is to make more cubs. But how do you do that?

Some tiger attractions use cruel methods to make sure they have enough cubs. They take young cubs away from their mothers before the cubs are ready to be independent. This makes the tigress angry and sad, so she has more cubs. Instead of having cubs two years apart like she might in the wild, a tigress in captivity might have cubs twice each year. This isn't healthy for the tigresses.

Once a roadside group has enough cubs for itself, it can sell the extras to other attractions. The rest are kept and displayed to tourists until they are old enough to have cubs of their own. Then the process starts again. This impacts these tigers for their entire lives. Once tiger cubs are hand-fed and kept in captivity, they can't be released into the wild. They haven't learned the skills to feed and protect themselves.

Have you seen tigers in magic shows or circuses? Those tigers follow commands, sit on podiums and jump through hoops of fire. Companies say this helps tourists learn about tigers. But those tricks are not natural behaviors for tigers. In order to make tigers do a task, many are trained with whips, tight collars and other harmful equipment. Circus animals may train or perform during the day. That's not natural for a nocturnal animal.

After people protested, many shows stopped using large animals like tigers in their acts. But others resisted. In 2017, the Ringling Bros. and Barnum & Bailey Circus closed their circus in

A tiger jumps through hoops of fire at a circus.

the United States. But they did not send their
tigers to a safe and legal sanctuary. Instead,
the company asked permission to move the
fourteen tigers to a circus in Germany. In July
2020, commissioners in Las Vegas, Nevada,
approved a magic act featuring three tigers.

Organizers planned to hold the show in a parking lot tent near the airport.

Citizens around the world are asking their governments to pass stronger laws that will fix these problems permanently. You can be part of the solution, too!

THE RACE TO SAVE TIGERS

As you have read, protecting tigers is becoming a world emergency. Tigers will disappear if we don't act quickly. The good news is there are plenty of people working on the problem. All of their efforts combine to help save tigers and their habitats. This helps protect planet Earth, too.

People to the Rescue!

The International Union for Conservation of Nature (IUCN) is doing some important work to help save tigers. We've already talked about the IUCN Red List. But did you know that the IUCN works with more than 1,400 organizations in 160 countries? More than 18,000 experts study the planet and share ideas about how to improve the environment and protect animals.

Some of those experts focus on improving conditions for wild tigers. Their members train park staff to protect tigers from poachers. Some members invented a special fence that helps farm animals and tigers coexist by keeping them away from each other. The good news is that the IUCN efforts to protect habitats are working. Between the years 2015 and 2021, there has been a 40 percent increase in the tiger populations at twelve projects they created.

The World Wildlife Fund (WWF) is a

global organization created in 1961. Its first office opened in Switzerland. The WWF was first concerned about the destruction of animal habitats. Since then, their mission has grown to include ways to save the planet. Now they also focus on climate, food, forests, fresh water,

A tiger cub follows her mother.

oceans and wildlife. They help communities find ways to conserve their natural resources. Governments all over the world are using their suggestions to protect animal habitats. Those efforts are working to save animals across the planet. In 2010, the International Tiger Forum was held in Russia. WWF members at that forum estimated the global wild tiger population to be 3,200. They set a goal to double the number by 2022, the designated "Year of the Tiger." Members called their goal TX2. That stands for "Tigers Times Two."

The Global Tiger Forum is based in India. It works with local governments to protect tigers and their prey. In 2015, the WWF and the Global Tiger Forum announced that the wild tiger population had increased for the first time in a hundred years. They counted 3,890 tigers. Sadly, in the years that followed, the number did not increase much more than that. The world still has work to do to help tigers survive.

The Animal Legal Defense Fund (ALDF) is a special group of lawyers for animals. They are based in the US, and they file lawsuits to stop people from illegally owning or mistreating wildlife, including tigers. They push federal and state governments to enforce laws already in place. In 2017, they sued an aquarium in Houston, Texas, for keeping four white tigers locked in indoor cages for thirteen years. The tigers, Nero, Marina, Coral and Reef, had never been outdoors.

In 2018, a court ruled that the tigers were protected by the 1973 Endangered Species Act. Under this law, Houston's Downtown

Aquarium was ordered to stop mistreating them. The aquarium worked with a licensed zoo to build a 3,500-square-foot outdoor habitat for the tigers. It has waterfalls, a pool for swimming and lots of plants. Most important, it has lots of sunlight. Even so, the space is small for four tigers. In comparison, a tiger habitat at the San Diego Zoo is 226,00 square feet for six tigers. That's almost sixty-five times larger than the Downtown Aquarium's tiger habitat!

The Association of Zoos and Aquariums (AZA) is an organization that monitors the zoos and aquariums that we love. You've probably visited a public zoo with family or on a school trip. Those zoos are not like roadside zoos. Our tax dollars and private donations help support many of them. It is estimated that only about one out of every twenty tigers held in captivity live in legal, licensed zoos. To prove they meet high standards for taking care of tigers and other animals, public zoos become members of the AZA. They have to be inspected every five years to keep their status. Top zoos like the Smithsonian National Zoo in Washington, DC, and the Henry Doorly Zoo and Aquarium in Omaha, Nebraska, are AZA members. That means you can be confident that they treat their tigers and other animals well. These zoos are safe for the animals and for kids like you.

The Global Federation of Animal Sanctuaries (GFAS) is a group that publishes a list of sanctuaries around the world that rescues tigers. Sanctuaries are excellent places to start if you want to learn about tigers. They do not buy or sell tigers. They don't force them to be on display for the public. Sanctuaries are places where injured and rescued tigers can live safely for the rest of their lives.

The federation lists more than two hundred approved animal sanctuaries in the world. But only thirteen are approved to take care of big cats like tigers. Most are in the United States. These sanctuaries provide large habitats, plenty of nutrition and clean water for tigers. They have rules in place to keep the tigers and the staff safe at all times. Veterinarians work to improve and protect the tigers' health.

The GFAS is not the only place that makes sure sanctuaries are safe. If you need more options, you can check out the Big Cat Sanctuary Alliance, Tigers in America and the American Sanctuary Association, as well as your nearest AZA-certified zoo.

Governments Take Action

India Increases Its Tiger Population

In 2019, officials in India counted 774 more tigers than in 2015. It showed that conservation efforts were slowly working. How did that happen? The Global Tiger Forum worked with the Indian government to create special reserves where tigers could live without being hunted by humans.

In 2022, there were thirty-five approved reserves across India, and more are planned. The total land set aside for tigers is more than 15,000 square miles. That's big enough to hold ten copies of the state of Rhode Island. If you include the protected forests around them, the land is more than 28,000 square miles. The Indian government made sure activities in nearby forests didn't interfere with tigers' home

A male tiger blocks the road at Ranthambore Tiger Reserve.

ranges. They trained people to patrol the land, track what was happening there and eliminate illegal hunting. This gave the tigers the space and time they needed to increase their numbers.

Ranthambore National Park in India is a perfect example. It covers 154 square miles and is one of the best-known tiger reserves in the world. Ranthambore has plenty of space for tigers to hunt, roam or just hide from humans. There are bodies of water for drinking and swimming in on a hot day. In 1980, the land was declared a national park.

Ranthambore became famous because of a Bengal tigress named Machli. She was comfortable around humans and often posed for them. Machli was so popular she was featured in a documentary called *The World's Most Famous Tiger*. She appeared in magazines, newspapers and even a book. So many people loved Machli that, while she was alive, the park earned up to $10 million each year. When she died, she was given a traditional Hindu funeral. Her body was wrapped in white cloth and covered in flowers.

A Bengal tiger takes a big leap forward at a water hole.

Thailand's Wildlife Conservation and Protection Act

Remember the Tiger Temple that we read about in the last chapter? It turns out there were a lot more. In fact, the Thai government found hundreds of illegal sanctuaries selling tigers that they had bought from poachers.

To try to solve this problem, the government

created the Wildlife Conservation and Protection Act in 2019. It made poaching and hunting tigers illegal. Now you can only hurt a tiger if you are trying to save your life. Even breeding tigers without government permission or buying their body parts became illegal. The Thai government is serious about enforcing these laws. One year, they arrested thirty-six poachers who tried to hunt tigers on protected land. Despite the law, "tiger zoos," where people can feed and pose with tigers, remained open.

While more needs to be done to protect tigers in Thailand, the government's efforts are working. In August 2021, the Thai government announced they'd counted seventeen more Indochinese tigers in the wild. This brings the total number of wild tigers to 177. Progress might seem slow, but it's a good start.

The United States Government
Hears from Angry Citizens

Remember the Captive Wildlife Safety Act and the Endangered Species Act? It didn't do enough to protect tigers in captivity. In 2020, a television series named *Tiger King* became popular. It put a spotlight on illegal tiger sanctuaries and petting zoos. After receiving calls and letters from angry citizens, Congress felt pressure to vote on the Big Cat Public Safety Act. It would stop private citizens from owning tigers. The bill had more than two hundred sponsors. The House of Representatives voted to approve it on December 3, 2020. But there was a problem. The Senate refused to vote on the bill, so the president couldn't sign it into law.

This may sound discouraging, but there's still hope. Other efforts are underway in Congress to continue trying to help protect tigers and other endangered animals. This includes passing stronger laws and getting the US Senate to vote on the Big Cat Safety Act. It's more important than ever to let your elected representatives at the state and federal level know that citizens like you want them to protect tigers and other big cats like other countries are doing.

Teaching Kids How to Protect Tigers

Fateh Singh Rathore was one of the most famous tiger protectors in the world. When he was in his twenties, Fateh became a forest ranger in the Sariska shooting reserve in India. Back then, it was legal to hunt tigers. He learned to track animals from another ranger, who would look for prey that tigers had killed. He also learned about wildlife conservation.

When the tiger population began dropping, the Indian government created Project Tiger. Fateh wanted to help. The Sariska reserve was the obvious choice. It had a lot more tigers. Instead, Fateh wanted a challenge. He asked to work on Ranthambore.

As a forest ranger, Fateh was always in the forest with his team finding ways to improve the environment. Soon Rathambore became a place where tigers and their prey could thrive. Having a healthy home range kept tigers from going close to villages where they might be killed.

Now came the hard part. He needed to make the park bigger. First, he had to convince sixteen villages to move to new land. But the move wasn't just the people. He had to convince them to move their farm animals to new grazing land, too. Fateh needed their old grazing land for the deer and gazelles that the tigers could hunt for food.

To do this, he used the resources of Project Tiger to give the villagers better land than they had before. It required patience. Poaching tigers continued, but Fateh was smart. He realized that villagers were hunting to support their families. Instead of arresting them, he gave them jobs.

Ranthambore became the best example of how to create a conservation program that works. But Fateh's efforts didn't stop there.

In 2000, the Kids for Tigers Festival was created by the Sanctuary Nature Foundation in India. More than 25,000 people attended. The festival taught children how to protect nature. The foundation believed that tiger numbers would bounce back if nature was allowed to heal itself. Fateh brought hundreds of children from nearby villages to learn to be forest protectors. One of those children was Govardhan Meena.

Years later, Govardhan has turned what he learned about tigers into action. He uses stories and films to show the villages how important tigers are for the environment. He connects villagers with park rangers to get jobs and to heal any conflicts if a tiger gets too close to a village. He and other villagers rescue injured tigers. He's now known as the Pied Piper of Ranthambore. And just like Fateh Singh Rathore, Govardhan Meena takes children through the forests in Ranthambore. So far he's hosted 15,000 children from forty-five forest villages.

Both men's actions show how one person can make a huge difference.

The Tiger Scouts of Bangladesh

Still wondering what a kid can do? Plenty.

In Bangladesh, a place called the Sundarbans is one of the largest mangrove forests in the world. It is also the only mangrove habitat left for Bengal tigers. But rising seas from climate change might destroy it.

An organization called WildTeam Bangladesh trains Bangladeshi students to be tiger scouts. The students learn how to teach other kids about tigers and their habitats and about the dangers of climate change.

Remember the IUCN, the organization we talked about earlier in this book? They partnered with the Wildlife Trust of India to create a conservation program. Because the Sundarbans shares a border with India, the program brought the tiger scouts from Bangladesh to meet with

Indian students who also cared about tigers. The students shared what they knew about protecting habitats and what might happen if tigers disappeared. The tiger scouts were so knowledgeable, they helped convince teachers and people from the forest department to take action as well. Now the small tiger population in the Sundarbans has increased by 26 percent.

Your Turn!

As you can see, the race to save tigers is on. But it will go faster if we all work together. Remember, a tiny drop of water can make a big ripple in a pond. You CAN make a difference!

FUN FACTS ABOUT TIGERS

1. Tigers are part of the big cat genus *Panthera*. Other animals in this category include lions, leopards, jaguars and snow leopards. Tigers are the only member of this group that have stripes.

2. Tigers share 96 percent of their DNA with domestic cats that people adopt as pets.

3. The oldest known fossil of a tiger is almost 2 million years old. It was found in China.

4. The scientific name for the end of a

tiger's whiskers is proprioceptor. That's a Latin word meaning "self receiver."

5. The collective name for a group of tigers is a streak or an ambush.

6. Tigers are so strong, one swipe of their paw could break your bones.

7. Tigers can mate with lions. If the mother is a tiger, the cub is called a liger. If the father is a tiger, the cub is called a tigon.

8. If a tiger licks its wound, the saliva in its mouth works like an antiseptic. It helps prevent infection.

9. Tiger urine smells like buttered popcorn.

10. International Tiger Day is celebrated on July 29 of each year. It was created in 2010 at the International Tiger Forum in Russia to help raise public awareness.

11. The stripes on a tiger's head are similar

to the Chinese character for "king."

12. 2022 was the Chinese Year of the Tiger. People born in the year of the tiger are considered natural leaders who believe in justice. The tiger is the third symbol in the Chinese zodiac. There are twelve animals in the zodiac, so each animal is celebrated every twelve years.

HOW YOU CAN HELP SAVE THE TIGERS

Now that you know more about tigers, you're probably thinking of great ways you can help them. Here are some to get you started and help turn your ideas into positive change for the future!

1. Don't buy clothes, jewelry or other items made from tigers, and ask adults to do the same. Buying these items encourages more illegal poaching. If there is no market for tiger parts, poachers will have to find another way to make money.

2. Don't visit roadside zoos. These zoos are owned by private citizens and do not go through the same inspection process as regular zoos to make sure the tigers are safe. If you want to visit a zoo that meets high standards, you can find a list at AZA.org. Ask those zoos how you can help them with their mission to protect animals, including tigers. Many have summer camps, nature walks or other kid-friendly programs. Can't travel to a licensed zoo? No problem! Some zoos, like the San Diego Zoo or the Indianapolis Zoo, have live tiger webcams on the internet. There's also a tiger webcam at the Edinburgh Zoo in Scotland. The cameras let you see what their tigers are doing even if you live far away.

3. Hold a fundraiser on International Tiger Day (July 29). You can have a yard sale or a bake sale, or open a lemonade stand. Donate the money you make from those to organizations working to help tigers. The money you raise would help them carry out their important work. Ask people to donate to your favorite organization, too. More resources will help these organizations save more tigers.

4. Check your food labels carefully. Many packaged foods are made with palm oil. If you can, buy foods that don't have palm oil. If they do, look for ones that carry the Roundtable on Sustainable Palm Oil (RSPO) logo. That means the oil palm trees were grown in a way that does not harm rainforests or animal habitats. If you can't find them, ask your store to carry those products. That's a good way to help companies that are acting responsibly. Eat fresh foods if you can. Natural foods don't contain palm oil and are good for your body. By cooking with fresh ingredients, you can help tigers and yourself at the same time.

5. Recycle. Many times, climate change and deforestation happen when companies make products to replace the things consumers throw away. It's a waste of resources. Recycling finds new uses for paper and plastic and helps reduce how much garbage is thrown into natural habitats that animals need to survive. Recycling saves trees and the environ-

ment. Doing this helps save tiger habitats, too.

6. Did you know the Animal Legal Defense Fund has student chapters at law schools around the country? You may be just a kid, but those law students will have information, videos and suggestions for how to get involved with saving animals, including the tigers you care about.

7. Write letters to your members of Congress in the House of Representatives and in the Senate, and to your state representatives, too. Tell them to pass strong laws to protect tigers. Organize your friends, classmates and parents to write or email, too. There is strength in numbers. Elected officials pay close

attention to the letters and emails they get from the public.

8. Most important, tell friends and family members what you've learned about tigers and ask them to spread the word. Many people may not know what is happening or how to help. Let your voice be heard. Sometimes change happens with a single person. Maybe today, that person can be you.

Together we can all save the tigers and help their numbers increase!

ACKNOWLEDGMENTS

I would like to thank Ken, Alexis and Olivia for putting up with my endless hours of research and revision. You're the best!

REFERENCES

American Institute of Physics—Inside Science News Service. "The Secret of a Tiger's Roar." ScienceDaily. December 29, 2000. https://www.sciencedaily.com/releases /2000/12/001201152406.htm.

Animal Legal Defense Fund. "Challenging Landry's Treatment of Endangered Tigers." Last modified December 27, 2019. https://aldf.org/case/challenging -landrys-treatment-of-endangered-tigers.

Association of Zoos and Aquariums. "Tiger

Conservation." https://www.aza.org
/tiger-conservation.

Barnes, Simon. *Tiger!* New York: St. Martin's
Press, 1995.

Black, Riley. "Mother Tigers Pass Down
Territory to Their Daughters." *National
Geographic*. August 10, 2010. https:
//www.nationalgeographic.com/science
/article/mother-tigers-pass-down-territory
-to-their-daughters-2.

Cushing, Andrew. "Why Do Tigers Have
Different Stripe Patterns?" The
Conversation. November 23, 2020.
https://theconversation.com/why-do
-tigers-have-stripes-145223.

Dutfield, Scott. "Tiger Guide: Species Facts,
How They Hunt and Where to See in the
Wild." Discover Wildlife. July 29, 2021.

https://www.discoverwildlife.com
/animal-facts/mammals/facts-about
-tigers.

Gurung, Abhishek. "Amazing Facts About
Tigers You May Not Know." Ranthambore
National Park. March 8, 2018. https:
//www.ranthamborenationalpark.com
/blog/amazing-facts-about-tigers.

International Union for Conservation of
Nature. "IUCN Red List of Threatened
Species." https://www.iucn.org/resources
/conservation-tools/iucn-red-list-threat-
ened-species.

Kennedy, Kostya, editor, et al. *LIFE: Tigers:
The World's Most Extraordinary Animal.*
New York: Meredith Corporation, 2021.

Luo, Shu-Jin, and Xiao Xu. "Save the White

Tigers." *Scientific American.* October 16, 2014. https://www.scientificamerican .com/article/save-the-white-tigers.

Matias, Dani. "Census Finds Nearly 3,000 Tigers in India." National Public Radio. July 29, 2019. https://www.npr.org/2019 /07/29/746237332/census-finds-nearly -3-000-tigers-in-india.

Meacham, Cory J. *How the Tiger Lost Its Stripes: An Exploration into the Endangerment of a Species.* New York: Harcourt Brace, 1997.

National Geographic. "Siberian Tiger." September 10, 2010. https:// www.nationalgeographic.com /animals/mammals/facts/siberian-tiger.

Reuters staff. "China Postpones Lifting of Ban on Trade of Tiger and Rhino Parts." Reuters. November 12, 2018. https://

www.reuters.com/article/us-china-wildlife
-idUSKCN1NH0XH.

San Diego Zoo Wildlife Alliance. "Tiger:
Panthera Tigris." San Diego Zoo Wildlife
Alliance Animals and Plants. https:
//animals.sandiegozoo.org/animals
/tiger.

Save China's Tigers. "Reintroduction to
China." https://www.savechinastigers
.org/reintroduction.html.

Sewing, Joy. "Controversial White Tigers at
the Downtown Aquarium Get a New
$4M Back Yard." *Houston Chronicle.* Last
modified December 24, 2019. https:
//www.houstonchronicle.com/culture
/main/article/Tilman-s-tigers-get-a-fancy
-new-back-yard-14919946.php.

Simons, Marlise. "Face Masks Fool the

Bengal Tigers." *The New York Times.*
September 5, 1989. https://www.nytimes
.com/1989/09/05/science/face-masks
-fool-the-bengal-tigers.html.

Smithsonian's National Zoo and Conservation
Biology Institute. "Great Cats: Tiger."
https://nationalzoo.si.edu/animals/tiger.

Smithsonian's National Zoo and Conservation
Biology Institute. "Just How Big Are
Tigers? What Does Tiger Poop Look Like?
And More Tiger Facts." July 29, 2020.
https://nationalzoo.si.edu/animals
/news/how-big-are-tigers-and-more-tiger
-facts.

US Fish and Wildlife Service. "Captive
Wildlife Safety Act." US Fish and Wildlife
Service Office of Law Enforcement. Last

modified February 14, 2013. https:
//www.fws.gov/le/captive-wildlife
-safety-act.html.

Wangchuk, Rinchen Norbu. "Meet 'Mr.
Ranthambhore,' The Forest Officer Who
Showed India How to Protect Tigers."
The Better India. August 5, 2020. https:
//www.thebetterindia.com/234711/how
-does-india-protect-tigers-fateh-singh
-rathore-ranthambhore-national-park
-rajasthan-india-nor41.

World Wildlife Fund. "Bold, Ambitious and
Visionary." About TX2. https://tigers
.panda.org/our_work/about_tx2_public.

World Wildlife Fund. "Tiger." https://www
.worldwildlife.org/species/tiger.

CHRISTINE TAYLOR-BUTLER is the author of more than ninety fiction and nonfiction books and articles for children. A graduate of MIT, she holds degrees in both civil engineering and art & design. She has served as a past literary awards judge for PEN America and for the Society of Midland Authors. She is an inaugural member of steaMG, an alliance of middle grade science fiction authors, and a contributor to STEM Tuesday. She lives in Kansas City with her husband, Ken, and a cat who thinks he's a dog.

Photo by Kecia Y. Stovall

You can visit Christine Taylor-Butler online at
ChristineTaylorButler.com
and follow her on Twitter
@ChristineTB

CHELSEA CLINTON is the author of the #1 *New York Times* bestseller *She Persisted: 13 American Women Who Changed the World*; *She Persisted Around the World: 13 Women Who Changed History*; *She Persisted in Sports: American Olympians Who Changed the Game*; *Don't Let Them Disappear: 12 Endangered Species Across the Globe*; *It's Your World: Get Informed, Get Inspired & Get Going!*; *Start Now!: You Can Make a Difference*; with Hillary Clinton, *Grandma's Gardens* and *The Book of Gutsy Women: Favorite Stories of Courage and Resilience*; and, with Devi Sridhar, *Governing Global Health: Who Runs the World and Why?* She is also the Vice Chair of the Clinton Foundation, where she works on many initiatives, including those that help empower the next generation of leaders. She lives in New York City with her husband, Marc, their children and their dog, Soren.

You can follow Chelsea Clinton on Twitter
@ChelseaClinton
or on Facebook at
Facebook.com/ChelseaClinton

DON'T MISS MORE BOOKS IN THE

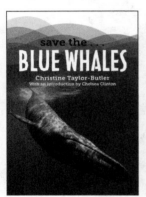

save the . . . SERIES!

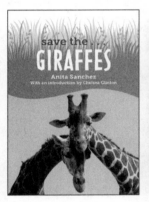

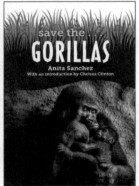

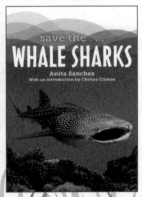